Cambridge Primary

Hodder Cambridge Primary

Maths

Activity Book

C

Foundation Stage

Ann and Paul Broadbent

HODDER
EDUCATION
AN HACHETTE UK COMPANY

Orders: please contact Bookpoint Ltd, 130 Park Drive, Milton Park, Abingdon, Oxon OX14 4SE. Telephone: (44) 01235 827720. Fax: (44) 01235 400401. Email: education@bookpoint.co.uk Lines are open from 9 a.m. to 5 p.m., Monday to Saturday, with a 24-hour message answering service. You can also order through our website: www.hoddereducation.com

© Ann Broadbent and Paul Broadbent 2018

First published in 2018

This edition published in 2018 by Hodder Education,
An Hachette UK Company
Carmelite House
50 Victoria Embankment
London EC4Y 0DZ
www.hoddereducation.co.uk

Impression number   10 9 8 7 6 5 4 3

Year   2022   2021   2020

Cover illustration by Steve Evans

Illustrations by Jeanne du Plessis, Vian Oelofsen

Typeset in FS Albert 17 pt by Lizette Watkiss

Printed in the United Kingdom

A catalogue record for this title is available from the British Library.

ISBN 978 1 5104 3184 3

MIX
Paper from
responsible sources
FSC™ C104740

Contents

Counting on from 10

⭐ Count on from 10.
Make these numbers with toys and cubes and count them.

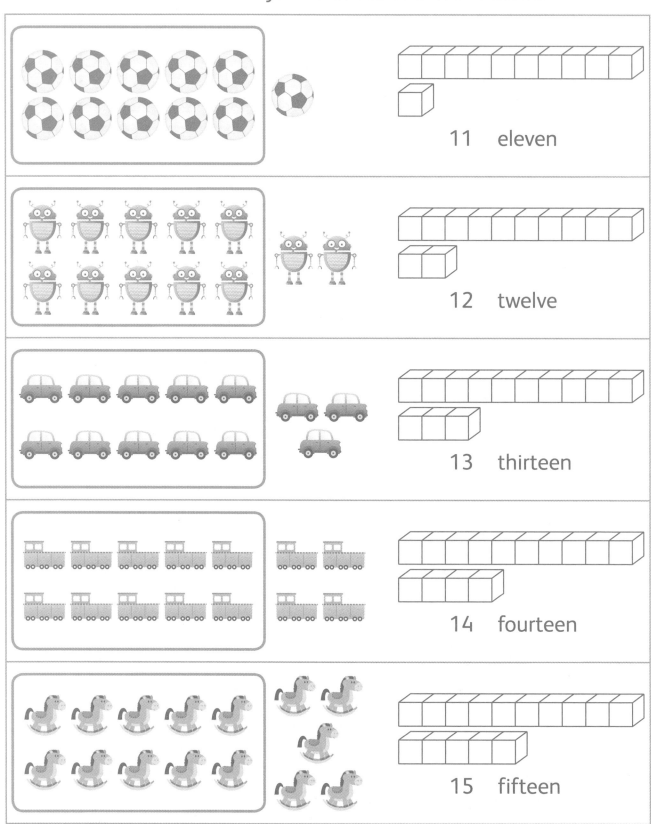

11   eleven

12   twelve

13   thirteen

14   fourteen

15   fifteen

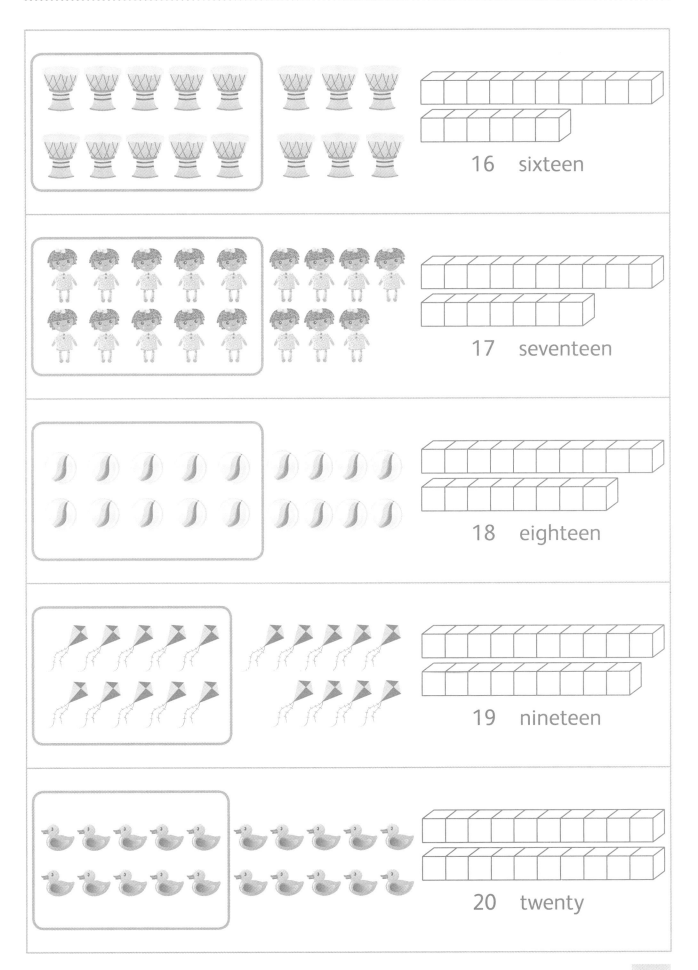

16   sixteen

17   seventeen

18   eighteen

19   nineteen

20   twenty

Counting to 20

Practise counting forwards and backwards on a number track.
Follow the numbers with your finger.

| 0 | 1 | 2 | 3 | 4 | 5 | 6 | 7 | 8 | 9 | 10 | 11 | 12 | 13 | 14 | 15 | 16 | 17 | 18 | 19 | 20 |

⭐ Draw lines to join each set of frogs to the correct number.

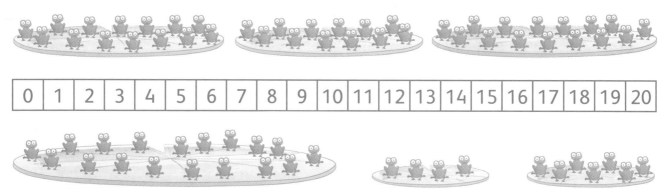

| 0 | 1 | 2 | 3 | 4 | 5 | 6 | 7 | 8 | 9 | 10 | 11 | 12 | 13 | 14 | 15 | 16 | 17 | 18 | 19 | 20 |

⭐ Draw 10 more fish in the pond.

There are ☐ fish altogether.

Numbers to 20

⭐ Write these numbers.

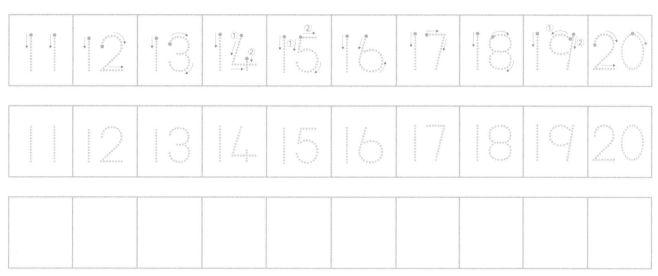

⭐ Count these shells. Write the numbers.

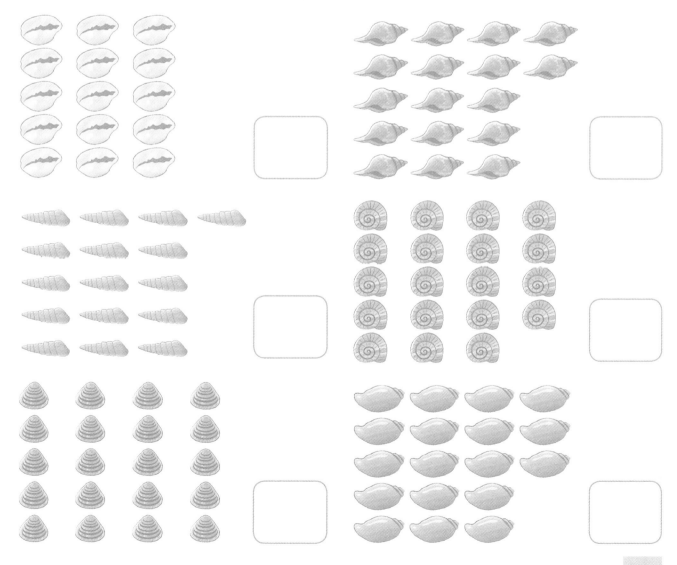

Counting groups

Point to each group and count the objects.

How many leaves are in each group?

How many groups of leaves are there?

How many leaves are there altogether?

 Count these in groups. Write how many leaves there are altogether.

Total: [ ]

Total: [ ]

Total: [ ]

Total: [ ]

Total: [ ]

⭐ Draw 2 apples on each tree. Count the total number of apples.

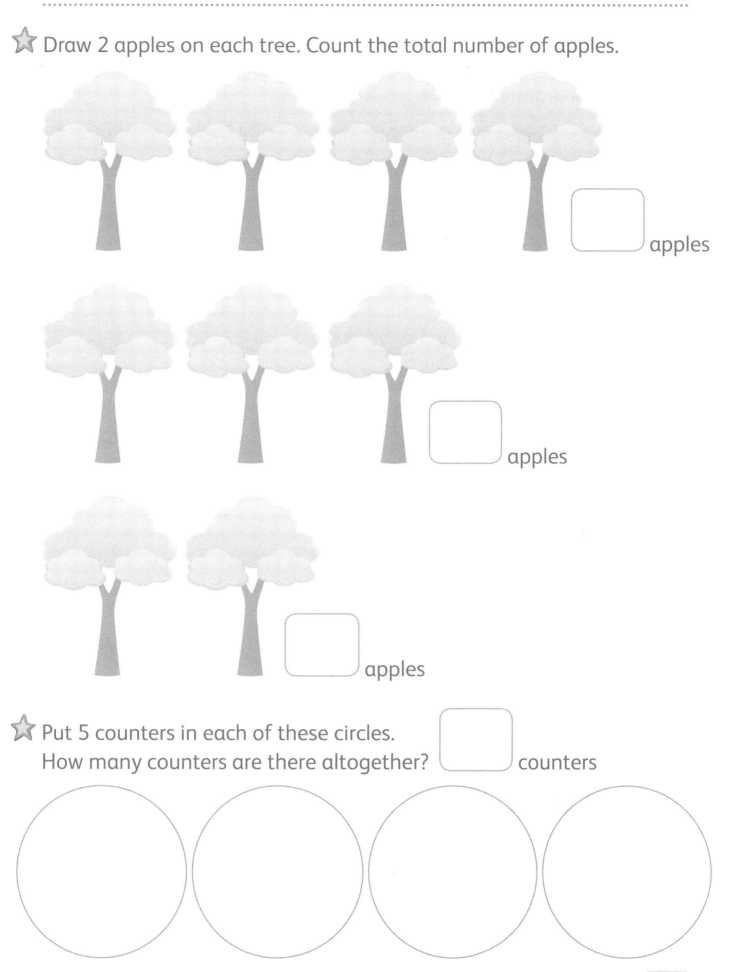

apples

apples

apples

⭐ Put 5 counters in each of these circles.
How many counters are there altogether? counters

Doubles

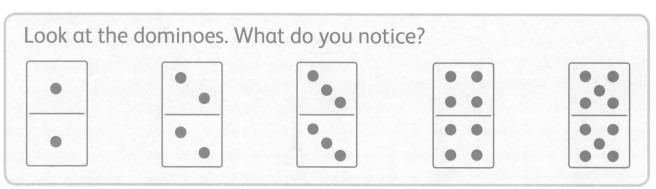

Look at the dominoes. What do you notice?

⭐ Use cubes to make each of these. Draw the cubes and write the total.

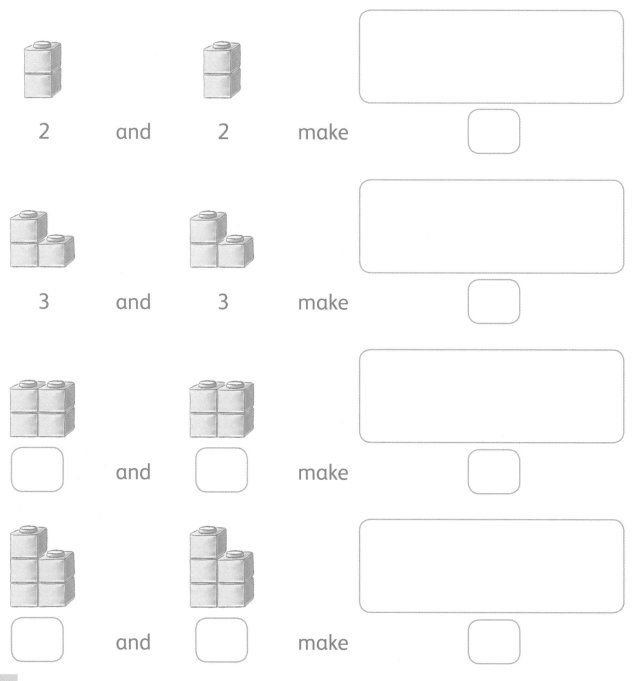

2 and 2 make

3 and 3 make

[ ] and [ ] make

[ ] and [ ] make

Sharing

⭐ Place counters in each box and count them.
Share the counters equally among the circles and write the numbers.

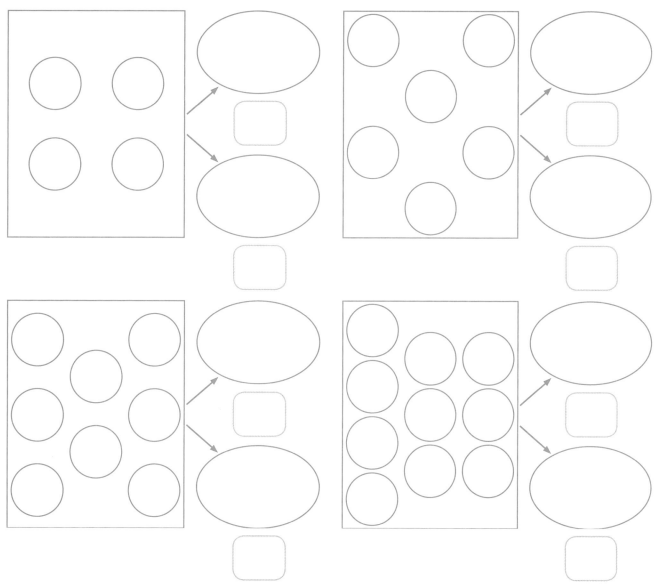

⭐ How many cakes can Kim and Mira have if they share each tray?

FRESH
CUPCAKES

Making totals

Look at the cubes. They show that 6 add 3 is equal to 9.

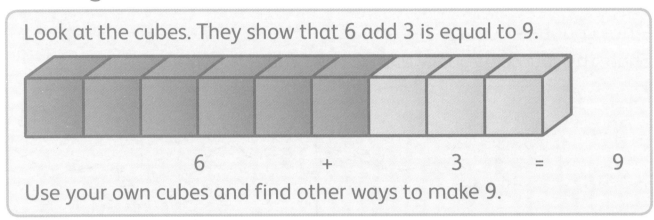

| 6 | + | 3 | = | 9 |

Use your own cubes and find other ways to make 9.

⭐ Use cubes to make each total in different ways.
Write the missing numbers.

| Make **8** | Make **7** | Make **6** |

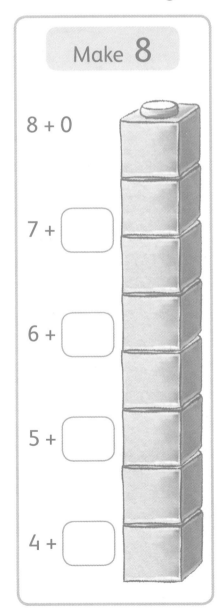

8 + 0

7 + ☐

6 + ☐

5 + ☐

4 + ☐

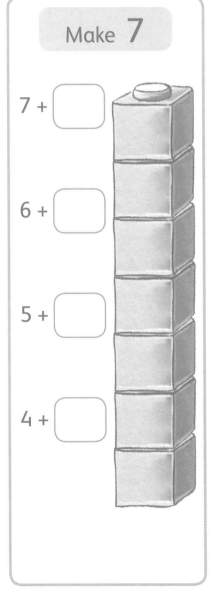

7 + ☐

6 + ☐

5 + ☐

4 + ☐

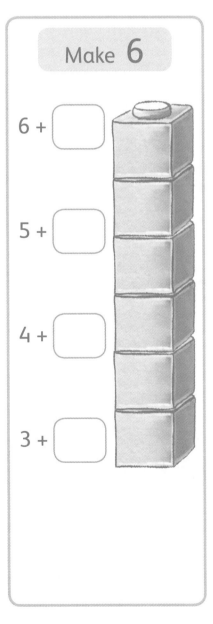

6 + ☐

5 + ☐

4 + ☐

3 + ☐

Join the additions to their matching answers.

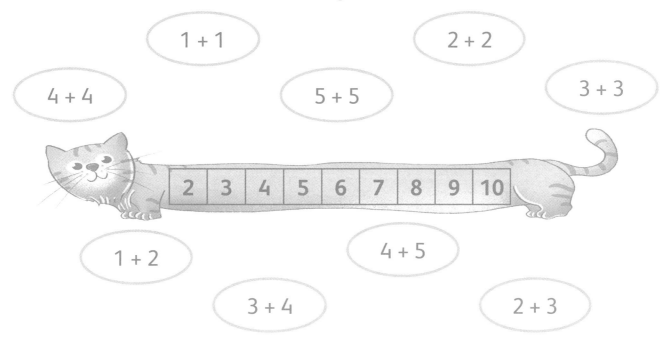

1 + 1

2 + 2

4 + 4

5 + 5

3 + 3

| 2 | 3 | 4 | 5 | 6 | 7 | 8 | 9 | 10 |

1 + 2

4 + 5

3 + 4

2 + 3

 Colour in all the additions on this grid that total 10.

6 + 5	5 + 5	3 + 7	9 + 1	4 + 4
7 + 2	2 + 8	3 + 4	4 + 6	9 + 0
4 + 5	7 + 3	10 + 0	6 + 4	2 + 7
7 + 1	0 + 9	6 + 3	8 + 2	2 + 6
8 + 3	2 + 7	1 + 8	0 + 10	3 + 5

What number can you see?

Counting on

Use a number track to help you add. Start at 5 and count on 3.

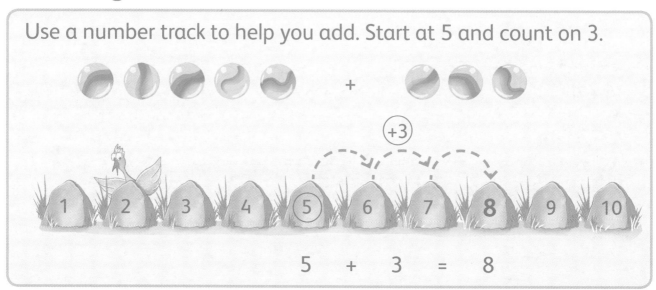

$$5 + 3 = 8$$

Play this game. You will need a number spinner.

How to play:
- Each player chooses a runner and spins the spinner twice.
- Place a counter on the first spin number.
- Move the counter as you count on the second spin number.
- Follow any instructions on the track where you land each time.
- The winner is the runner that has gone furthest after two spins.

Make a spinner like this.

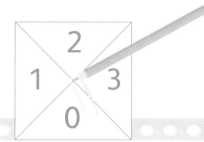

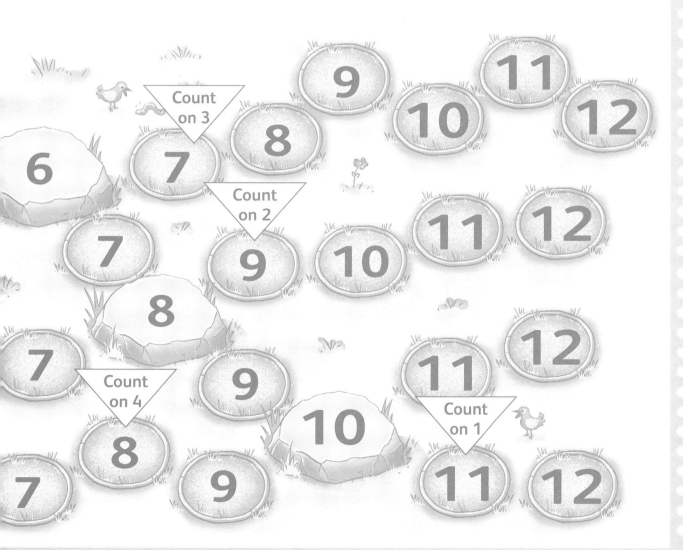

Addition stories

 Add the spots on each domino.

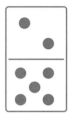 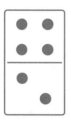 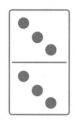 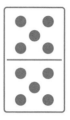 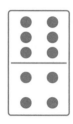 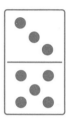

2 and 5   3 and 3   4 and 2   5 and 5   6 and 4   3 and 5

 Choose a number to complete each fact.
Colour in the counters to help you.

4 + ⬚ = 7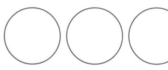

5 + ⬚ = 6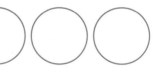

⬚ + 3 = 5

⬚ + 3 = 8

1 + ⬚ = 5

2 + ⬚ = 8

Read this addition story.

4 fish are swimming in the tank.
1 crab is on the sand in the tank.
There are 5 sea creatures altogether.

⭐ Use these pictures to make addition stories.

1 + 2 = ☐

4 + 3 = ☐

4 + 6 = ☐

5 + 2 = ☐

Finding the difference

Compare these towers.
Count on from 4 to 7.
The difference is 3.

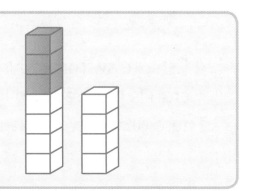

 Colour the cubes to show the difference each time. Write the number.

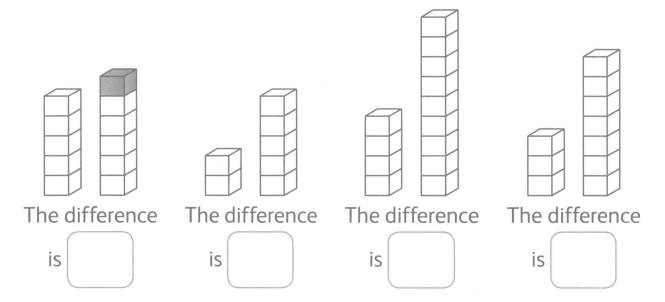

The difference
is [ ]

The difference
is [ ]

The difference
is [ ]

The difference
is [ ]

☆ Draw lines to join the pairs of numbers with a difference of 4.

1

4

2

3

5

9

5

7

8

6

Counting back

Count back from 10 to 0.

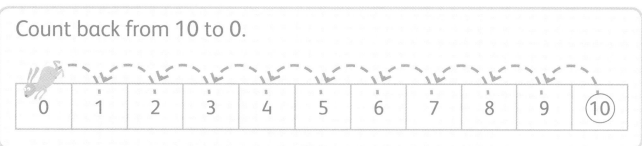

| 0 | 1 | 2 | 3 | 4 | 5 | 6 | 7 | 8 | 9 | ⑩ |

 Draw the jumps to count back. Circle the number that you finish on.

Count back 2.

| 0 | 1 | 2 | 3 | 4 | 5 | 6 | 7 | 8 | 9 | 10 |

6 take away 2 is ☐

Count back 4.

| 0 | 1 | 2 | 3 | 4 | 5 | 6 | 7 | 8 | 9 | 10 |

7 take away 4 is ☐

Count back 3.

| 0 | 1 | 2 | 3 | 4 | 5 | 6 | 7 | 8 | 9 | 10 |

9 take away 3 is ☐

Count back 5.

| 0 | 1 | 2 | 3 | 4 | 5 | 6 | 7 | 8 | 9 | 10 |

8 take away 5 is ☐

More counting back

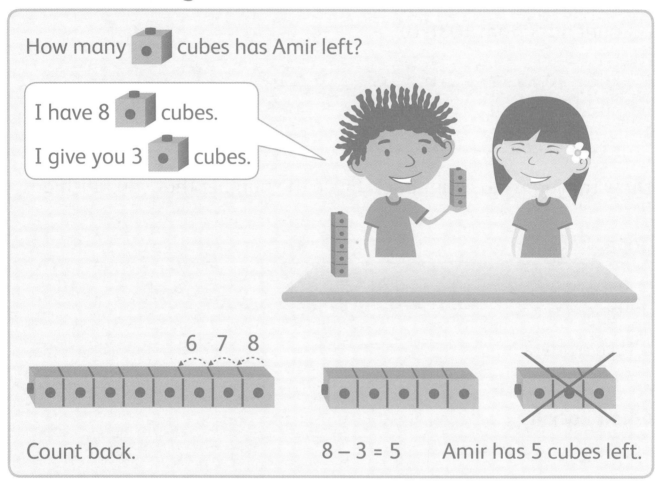

How many 🔲 cubes has Amir left?

I have 8 🔲 cubes.

I give you 3 🔲 cubes.

6 7 8

Count back.   8 − 3 = 5   Amir has 5 cubes left.

⭐ Count back. Cross out the cubes to find what is left.
Write the number.

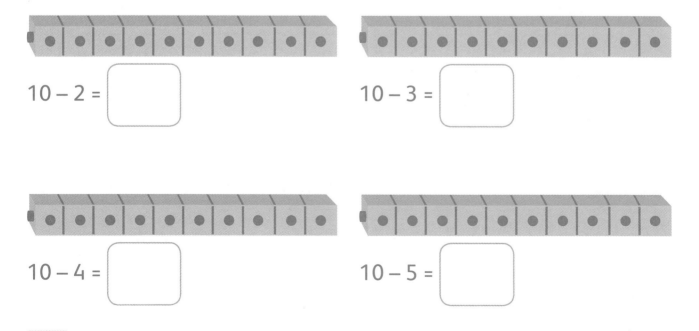

10 − 2 = ☐   10 − 3 = ☐

10 − 4 = ☐   10 − 5 = ☐

Hiding objects

There are 5 eggs in this nest. We can still see 2 eggs.
How many eggs are hidden?

⭐ There are 6 eggs in each nest.
How many eggs are hidden?

  eggs
are hidden.

 eggs
are hidden.

 eggs
are hidden.

eggs
are hidden.

⭐ Draw some eggs in this nest. Colour 1 egg green.
How many eggs are not green?

Measuring length

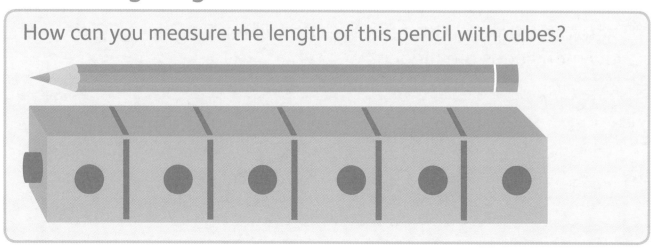

How can you measure the length of this pencil with cubes?

⭐ Measure the length of these ribbons using cubes or counters.

8

Cubes, sticks and paper clips are used to measure the length of this book.

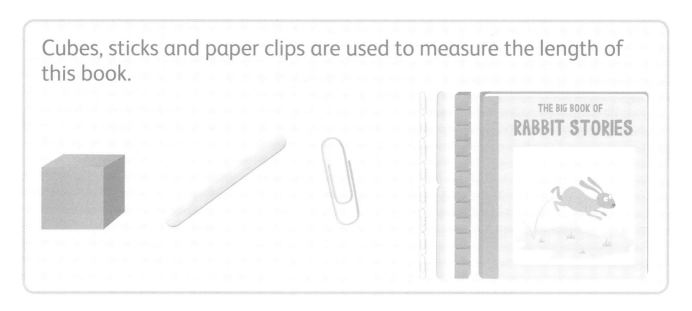

Write the length of the book in cubes, sticks and paper clips.

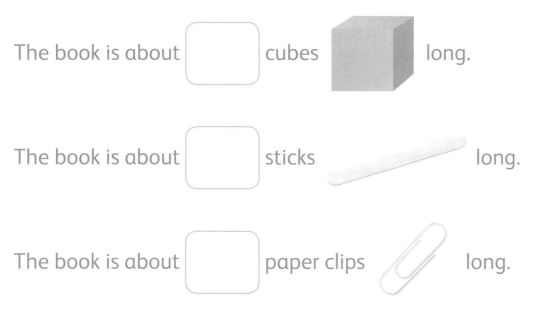

The book is about ☐ cubes ▣ long.

The book is about ☐ sticks ╱ long.

The book is about ☐ paper clips ⬭ long.

⭐ Use different objects to measure the length of a table.
Which object is the easiest to use?

Weight and capacity

Weight

We use the words **heavy** and **light** when we talk about weight.

Capacity

The word **capacity** means how much something holds.

⭐ Which objects are **heavy** and which are **light**?
Circle the heavier object in each group.

⭐ Colour red the largest container. Circle the smallest container.
Tick the container that is full.

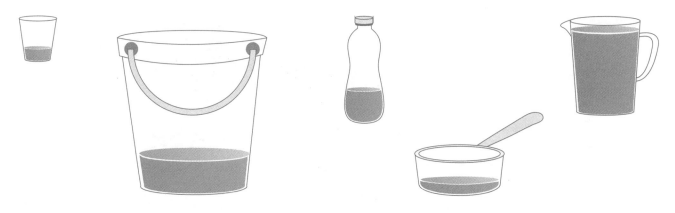

Time

> The long hand is pointing to the 12.
> The short hand is pointing to the 3.
> This shows 3 o'clock.

⭐ Write these times.

_____ o'clock

_____ o'clock

_____ o'clock

_____ o'clock

_____ o'clock

_____ o'clock

_____ o'clock

_____ o'clock

⭐ Colour all the clocks that show 5 o'clock.

Flat shapes

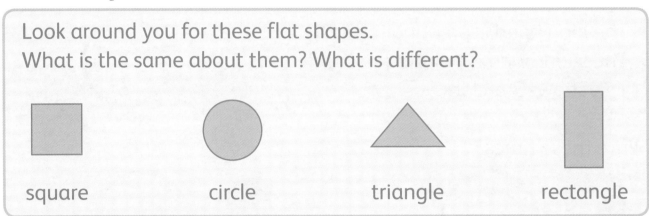

Look around you for these flat shapes.
What is the same about them? What is different?

square   circle   triangle   rectangle

⭐ How many of these shapes can you see in the picture?

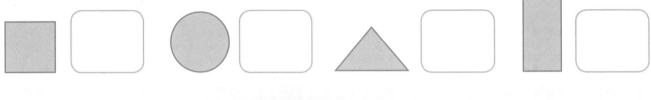

⭐ Draw the next three beads in each row.
Colour the shapes to make patterns.

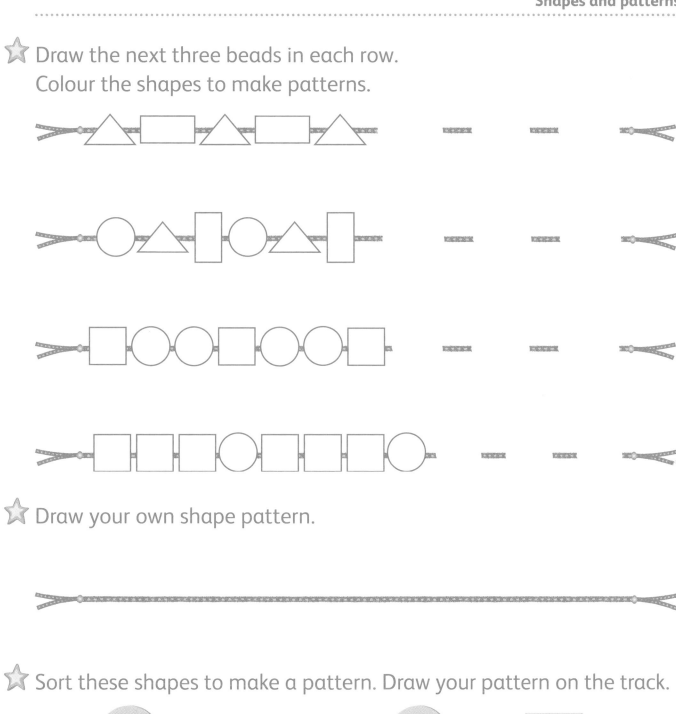

⭐ Draw your own shape pattern.

⭐ Sort these shapes to make a pattern. Draw your pattern on the track.

Solid shapes

Look around you for these solid shapes. Try to remember their names.

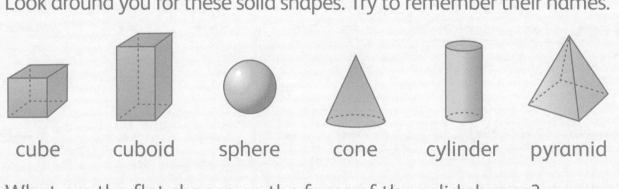

cube   cuboid   sphere   cone   cylinder   pyramid

What are the flat shapes on the faces of the solid shapes?

⭐ Tick the shapes that have some faces that are square.

⭐ Cross out the shape that is different in each set.

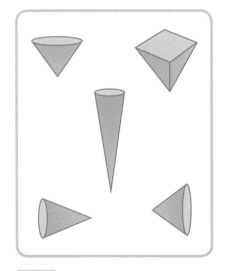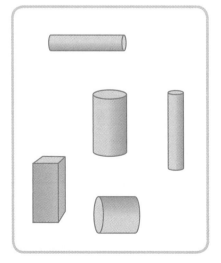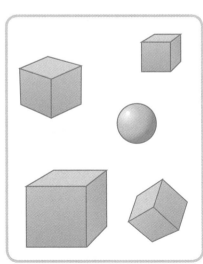

Symmetry

These shapes are symmetrical.

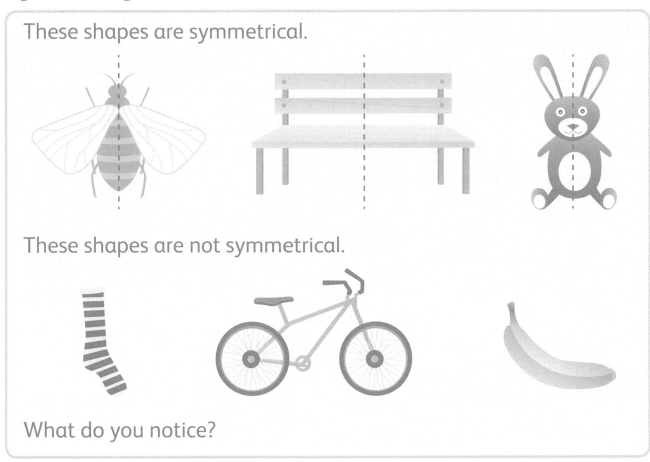

These shapes are not symmetrical.

What do you notice?

⭐ Colour the symmetrical shapes.

What can you remember?

⭐ Read these numbers.
Cover a number with a counter.
Can you say which number is covered?

1	2	3	4	5
6	7	8	9	10
11	12	13	14	15
16	17	18	19	20

⭐ Put 4 counters in each of these circles.
How many counters are there altogether?

◯　　　◯　　　◯　　　▢ counters

⭐ Draw jumps on the number tracks to help you add these objects.
Write the answers.

2　+　▢　=　▢

1	2	3	4	5	6	7	8	9	10

3　+　▢　=　▢

1	2	3	4	5	6	7	8	9	10

▢　+　▢　=　▢

1	2	3	4	5	6	7	8	9	10

▢　+　▢　=　▢

1	2	3	4	5	6	7	8	9	10

Cross out the cubes to find the answers.

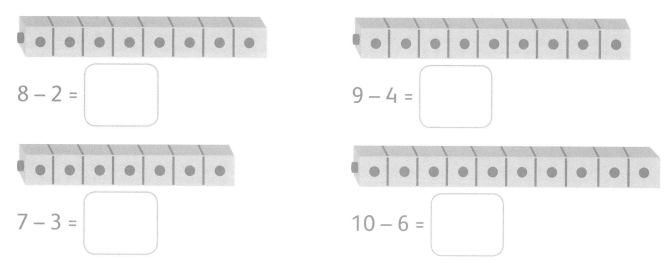

8 − 2 = ☐

9 − 4 = ☐

7 − 3 = ☐

10 − 6 = ☐

 Draw a line to join cube sticks of the same length.

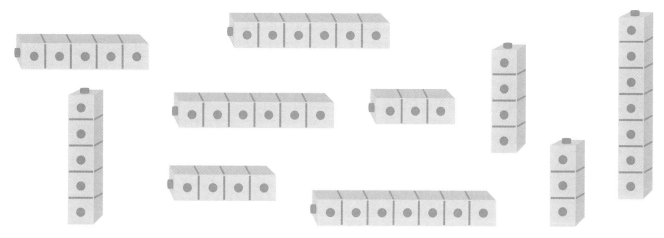

 Circle the shape in each row that is not the same.
How is it different?

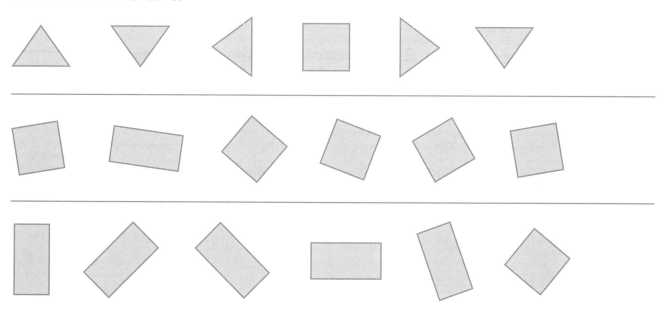

Self-assessment

Colour the stars to show what you can do!

Numbers to 20	I can read and write the numbers to 20.	☆
	I can describe where a number to 20 is on a number track.	☆
	I can count up to 20 objects and say the total.	☆
Grouping and sharing	I can put objects into equal groups and count the groups and totals.	☆
	I can double and halve small sets of objects.	☆
	I can share objects between two people.	☆
Addition	I can use a number track to add by counting on.	☆
	I can put two groups of objects together and count on from the largest group to find the total.	☆
	I can make up addition stories.	☆
Subtraction	I can compare two cube towers and say how many more or less there are.	☆
	I can use a number track to take away by counting back.	☆
	I can work out the missing number of objects.	☆
Measures and time	I can measure the length of a table using cubes.	☆
	I can use a balance to find out which objects are heavier than others.	☆
	I can compare the capacity of containers and say which is the most full.	☆
	I know some times on a clock.	☆
Shapes and patterns	I can describe flat shapes and what makes them the same or different from others.	☆
	I can make repeating patterns with shapes.	☆
	I can describe the faces of different solid shapes.	☆
	I can show a symmetrical shape.	☆